EFFECTS OF
CHANGING CLIMATE
ON WEATHER
AND HUMAN ACTIVITIES

EFFECTS OF
CHANGING CLIMATE
ON WEATHER
AND HUMAN ACTIVITIES

Kevin E. Trenberth, Kathleen Miller,
Linda Mearns and Steven Rhodes
National Center for Atmospheric Research
Boulder, Colorado

UNIVERSITY SCIENCE BOOKS
SAUSALITO, CALIFORNIA

University Science Books
55D Gate Five Road
Sausalito, CA 94965
Fax: (415) 332-5393
www.uscibooks.com

Scientific director: Tom M.L. Wigley
Managing editor: Lucy Warner
Editor: Carol Rasmussen
Art and design: NCAR Image and Design Services
Cover design and composition: Craig Malone
Cover photo by Mickey Glantz

The cover photograph of the effects of drought on a farm in eastern Colorado in 1977 is prototypical of scenes in the 1930s during the "dust bowl" era. The risk of such droughts with global warming increases owing to increased drying of the landscape.

This book is printed on acid-free paper.

Library of Congress Cataloging-in-Publication Data

Effects of changing climate on weather and human activities / Kevin Trenberth ... [et al.].
    p. cm. – (The global change instruction program)
    Includes bibliographical references and index.
    ISBN 1-891389-14-9 (softcover : alk. paper)
    1. Climatic changes. 2. Weather. 3. Human beings–Effect of climate on. I. Trenberth, Kevin E. II. Series.

QC981.8.C5 E44 2000
304.2'5–dc21

                                                            00-023978

Printed in the United State of America
10 9 8 7 6 5 4 3 2 1

A Note on the Global Change Instruction Program

This series has been designed by college professors to fill an urgent need for interdisciplinary materials on global change. These materials are aimed at undergraduate students not majoring in science. The modular materials can be integrated into a number of existing courses—in earth sciences, biology, physics, astronomy, chemistry, meteorology, and the social sciences. They are written to capture the interest of the student who has little grounding in math and the technical aspects of science but whose intellectual curiosity is piqued by concern for the environment.

For a complete list of modules available in the Global Change Instruction Program, contact University Science Books, Sausalito, California, univscibks@igc.org. Information is also available on the World Wide Web at http://www.uscibooks.com/globdir.htm or http://www.ucar.edu/communications/gcip/.

Contents

Preface

It is now widely recognized that human activities are transforming the global environment. In the time it has taken for this book to come to fruition and be published, the evidence for climate change and its disruption of societal activities has become stronger. In the first 11 months of 1998, there were major floods in China, Peru, and California, enormous damage from Hurricane Mitch in Central America, record-breaking heat waves in Texas, and extensive drought and fires in Indonesia; weather-related property losses were estimated at over $89 billion, tens of thousands of lives were lost, and hundreds of thousands of people were displaced. This greatly exceeds damage estimates for any other year. The environment was ravaged in many parts of the globe. Many of these losses were caused by weird weather associated with the biggest El Niño on record in 1997–98, and they were probably exacerbated by global warming: the human-induced climate change arising from increasing carbon dioxide and other heat-trapping gasses in the atmosphere. The climate is changing, and human activities are now part of the cause. But how does a climate change manifest itself in day-to-day weather?

This book approaches the topic by explaining distinctions between weather and climate and how the rich natural variety of weather phenomena can be systematically influenced by climate. Appreciating how the atmosphere, where the weather occurs, interacts with the oceans, the land surface and its vegetation, and land and sea ice within the climate system is a key to understanding how influences external to this system can cause change. One of those influences is the effect of human activities, especially those that change the atmospheric composition with long-lived greenhouse gases.

Climate fluctuates naturally on very long time scales (thousands of years), and it is the rapidity of the projected changes that are a major source of concern. The possible impacts of the projected changes and how society has responded in the past and can in the future are also described. Everyone will be affected one way or another. So this is an important topic, yet it is one about which a certain amount of disinformation exists. Therefore it is as well to understand the issues in climate change and how these may affect each and every one of us. What we should do about the threats, given the uncertainties, is very much a choice that depends upon values, such as how much we should be stewards for the planet and its finite resources for the future generations. Many people favor a precautionary principle, "better safe than sorry," and err on the side of taking actions to prevent a problem that might not be as bad as feared. This book helps provide the knowledge and enlightenment desirable to ensure that the debate about this can be a public one and carried out by people who are well informed.

Kevin E Trenberth

Acknowledgments

This instructional module has been produced by the Global Change Instruction Program of the University Corporation for Atmospheric Research, with support from the National Science Foundation. Any opinions, findings, conclusions, or recommendations expressed in this publication are those of the authors and do not necessarily reflect the views of the National Science Foundation.

This project was supported, in part, by the
National Science Foundation
Opinions expressed are those of the authors
and not necessarily those of the Foundation

Introduction

We experience weather every day in all its wonderful variety. Most of the time it is familiar, yet it never repeats exactly. We also experience the changing seasons and associated changes in the kinds of weather. In summer, fine sunny days are interrupted by outbreaks of thunderstorms, which can be violent. Outside the tropics, as winter approaches the days get shorter, it gets colder, and the weather typically fluctuates from warm, fine spells to cooler and snowy conditions. These seasonal changes are the largest changes we experience at any given location. Because they arise in a well-understood way from the regular orbit of the Earth around the Sun, we expect them, we plan for them, and we even look forward to them. We readily and willingly plan (and possibly adapt) summer swimming outings or winter ski trips. Farmers plan their crops and harvests around their expectation of the seasonal cycle.

By comparison with this cycle, variations in the average weather from one year to the next are quite modest, as they are over decades or human lifetimes. Nevertheless, these variations can be very disruptive and expensive if we do not expect them and plan for them. For example, in summer in the central United States, the major drought in 1988 and the extensive heavy rainfalls and flooding in 1993 were at the extremes for summer weather in this region. (In the upper Mississippi Basin, rainfalls in May, June, and July changed from about 150 millimeters in 1988 to over 500 mm in 1993.) These two very different summers were the result of very different weather patterns. We assumed, before their occurrence, that the usual summertime mix of rain and sun would occur and that farmers' crops would flourish. Because this assumption was wrong, major economic losses occurred in both years and lives were disrupted.

These weather patterns and kinds of weather constitute a short-term climate variation or fluctuation. If they repeat or persist over prolonged periods, then they become a climate change. For instance, in parts of the Sahara Desert we now expect hot and dry conditions, unsuitable for human habitation, where we know that civilizations once flourished thousands of years ago. This is an example of a climate change.

How has the climate changed? What are the factors contributing to climate and therefore to possible change? How might climate change in the future? How does a change in climate alter the weather that we actually experience? How much certainty can we attach to any predictions? What do we do in the absence of predictability? Why are climate change and associated weather events important? What are the likely impacts on human endeavors and society and on natural-resource-based economic activities, such as agriculture? These are some of the questions we address in this module. Our discussion of impacts will focus on human activities. Although very important, the impacts of climate change on the natural environment and the unmanaged biosphere are not dealt with here. Some of these consequences are discussed further in the Global Change Instruction Program module *Biological Consequences of Global Climate Change*.

Many of these questions, although of considerable importance, unfortunately do not have simple answers. Also, many of the answers are not very satisfying. Because of the nature of the phenomena involved, many outcomes can only be stated in a statistical or probabilistic way.

We first need to distinguish between weather and climate. An important concept to grasp is how weather patterns and the kinds of weather that occur relate to climate. We refer to this relationship as the "weather machine" because of the way the weather helps drive the climate system. It is the sum of many weather phenomena that determines how the large-scale general circulation of the atmosphere (that is, the average three-dimensional structure of atmospheric motion) actually works; and it is the circulation that essentially defines climate. This intimate link between weather and climate provides a basis for understanding how weather events may change as the climate changes.

There are many very different weather phenomena that can take place under an unchanging climate, so a wide range of conditions occurs naturally. Consequently, even with a modest change in climate, many if not most of the same weather phenomena will still occur. Because of this large overlap between the weather events experienced before and after some climate change, it may be difficult to perceive such a change. Our perceptions are most likely to be colored not by the more common weather events but by extreme events. As climate changes, the frequencies of different weather events, particularly extremes, will change. It is these changes in extreme conditions that are most likely to be noticed.

We normally (and correctly) think of the fluctuations in the atmosphere from hour to hour or day to day as weather. Weather is described by such elements as temperature, air pressure, humidity, cloudiness, precipitation of various kinds, and winds. Weather occurs as a wide variety of phenomena ranging from small cumulus clouds to giant thunderstorms, from clear skies to extensive cloud decks, from gentle breezes to gales, from small wind gusts to tornadoes, from frost to heat waves, and from snow flurries to torrential rain. Many such phenomena occur as part of much larger-scale organized weather systems which consist, in middle latitudes, of cyclones (low pressure areas or systems) and anticyclones (high pressure systems),

and their associated warm and cold fronts. Tropical storms are organized, large-scale systems of intense low pressure that occur in low latitudes. If sufficiently intense these become hurricanes, which are also known as typhoons or tropical cyclones in other parts of the world. Weather systems develop, evolve, mature, and decay over periods of days to weeks. From a satellite's viewpoint, they appear as very large eddies, similar to the turbulent eddies that occur in streams and rivers, but on a much greater scale. Technically, they are indeed forms of turbulence in the atmosphere. They occur in great variety, but within certain bounds and over fairly short time frames.

Climate, on the other hand, can be thought of as the average or prevailing weather. The word is used more generally to encompass not only the average, but also the range and extremes of weather conditions, and where and how frequently various phenomena occur. Climate extends over a much longer period of time than weather and is usually specified for a certain geographical region. It has been said that climate is what we expect, but weather is what we get! Climate involves variations in which the atmosphere is influenced by and interacts with other parts of the climate system, the oceans, the land surface, and ice cover. Climate can change because of changes in any of these factors or if factors outside the Earth or beyond the climate system force it to change.

The Earth's climate has changed in the past and is expected to change in the future. We will experience these changes through the day-to-day weather. It is natural to want to ascribe a cause to any perceived unusual weather, and "climate change" is often espoused by the popular press as a possible cause. In some cases this inference may be correct—but proving it to be correct is exceedingly difficult. More often, extremes of weather occur simply as a manifestation of various interacting atmospheric processes. In other words, extremes are generally nothing more than examples of the tremendous natural variability that characterizes the atmosphere.

These considerations make it essential to understand and deal with the natural variability in the climate system. One way of thinking about the variability in the atmosphere is to consider the inherent natural variability as being in the realm of "weather," while systematic changes in the atmosphere that can be linked to a cause, such as interactions with the ocean or changes in atmospheric composition, are in the realm of climate.

For example, interactions between the atmosphere and the tropical Pacific Ocean result in the phenomenon known as El Niño, which is responsible for disruptions in weather patterns all over the world. Technically, El Niño is a warming of the eastern equatorial Pacific that occurs every two to six years and lasts for several seasons; it is a natural phenomenon and has occurred for thousands of years at least. It causes heavy rainfall along the western South American coast and southern part of the United States; drought or dry conditions in Australia, Indonesia, southeastern Asia (including the Indian subcontinent), parts of Africa, and northeast Brazil and Colombia; and unusual weather patterns in other parts of the world. It can be thought of as a short-term climatic phenomenon.

Other climate perturbations are more subtle and their effects on weather less obvious. Increases in heat-retaining gases called greenhouse gases, the best known of which is carbon dioxide, are currently causing the climate to warm because of human activities. In this case, the climate change is very gradual and should be noticeable only when the weather from one decade is compared with that of another. Even then, because of the background natural variability of the climate system, weather variations specifically attributable to human influences may be extremely difficult to identify.

While increasing greenhouse-gas concentrations cause global-mean warming, this does not mean that the globe will warm everywhere at once. An example is the Northern Hemisphere winter of December 1993 to February 1994. This winter was very cold and snowy, with many-more-than-normal winter storms in the northeastern part of the United States. How does this jibe with expectations of global warming?

The pattern of exceptionally wintry weather continued for several months, long enough to heighten interest in its apparent climate implications. However, as part of this pattern, there were often mild and sunny conditions in the western half of the United States and Canada, with above-average temperatures. Temperatures were substantially above average in parts of southeast Asia, northern Africa, the Mediterranean, and the Caribbean. The Northern Hemisphere as a whole was 0.2°C above the average for 1951 to 1980.

Extensive regions of above and below normal temperatures are the rule, not the exception, even in the presence of overall warmer conditions. A bout of below-average temperatures regionally may not be inconsistent with global warming, just as a bout of above-average temperatures may not indicate global warming.

In the following pages, a discussion is presented of how the climate may change and the reasons for possible changes. The primary reason for particular future climate change is the continuing influence of humans, especially through changes in atmospheric composition such as increases in greenhouse gases (notably carbon dioxide). We therefore pay particular attention to these effects and attempt to translate them into weather changes. A further issue is how these changes may in turn affect human activities. Accordingly, we consider how possible changes in climate and weather affect various economic sectors and human activities, and we discuss some steps that can be taken to soften the possible impacts.

I
Climate

The Climate System

The Earth's climate involves variations in a complex system in which the atmosphere interacts with many other parts (Figure 1). The other components of this climate system include the oceans, sea ice, and the land and its features. Important characteristics on land include vegetation, ecosystems, the total amount of living matter (or biomass), and the reflectivity of the surface (or albedo). Water is a central element of the climate system, and it appears in many forms: snow cover, land ice (including glaciers and the large ice sheets of Antarctica and Greenland), rivers, lakes, and surface and subsurface water.

Climate is also affected by forces outside this system: radiation from the Sun, the Earth's rotation, Sun-Earth geometry, and the Earth's slowly changing orbit (Figure 2). Over long

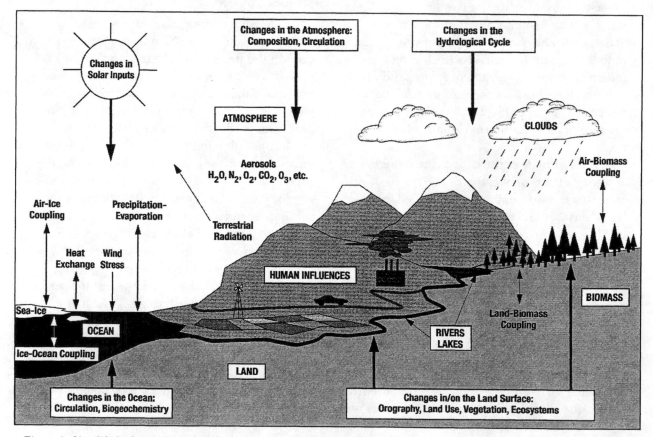

Figure 1. Simplified schematic view of the components of the global climate system and their interactions. Components of the climate system are indicated in bold type in boxes. Larger boxes at the top and bottom indicate the potential changes. Interactions are shown by the arrows.

periods of time, the physical and chemical makeup of the Earth's surface also changes. Continents drift, mountains develop and erode, the ocean floor and its basins shift, and, in addition to water vapor changes, the composition of the dry atmosphere also changes. These alterations, in turn, change the climate.

Radiation is measured in Watts or per unit area in Watts per meter squared (W/m^2). Averaged over day and night, as well as over all parts of the world, the solar radiation received at the top of the atmosphere is 342 W/m^2 or 175 PetaWatts (175,000,000,000,000,000 Watts). For comparison, a typical light bulb puts out 100 Watts, and a one-bar electric heater is 1,000 Watts.

Atmospheric composition is fundamental to the climate. Most of the atmosphere consists of nitrogen and oxygen (99% of dry air). Sunlight passes through these gases without being absorbed or reflected, so the gases have no climatic influence. The climate-relevant gases reside in the remaining 1% of dry air, together with water vapor. Some of these gases absorb a portion of the radiation leaving the Earth's surface and re-emit it from much higher and colder levels out to space. Such gases are known as greenhouse gases, because they trap heat and make the atmosphere substantially warmer than it would otherwise be, somewhat analogous to the effects of a greenhouse. This blanketing is known as the natural greenhouse effect. The main greenhouse gases are water vapor, which varies in amount from about 0 to 2%; carbon dioxide, which is about 0.04% of the atmosphere; and some other minor gases present in the atmosphere in much smaller quantities.

The greatest changes in the composition of the atmosphere are entirely natural and involve water in various phases in the atmosphere: as water vapor, clouds of liquid water and/or ice crystal clouds, and rain, snow, and hail. Other constituents of the atmosphere and the oceans can also change. A change in any of the climate system components, whether it is initiated inside or outside of the system, causes the Earth's climate to change.

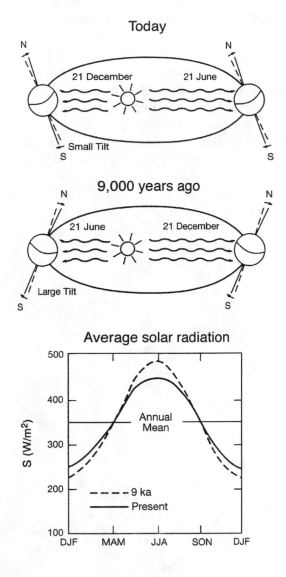

Figure 2. Top: The Earth's orbit around the Sun, illustrating the seasons in both current times and 9,000 years ago. Today the Earth is nearest the Sun in northern winter, and has an axial tilt of 23 1/2 degrees; in the past, the Earth was nearest the Sun in northern summer and tilted by 24 degrees. Bottom: Changes in average Northern Hemisphere solar radiation, in Watts per square meter, from 9,000 years ago (ka) to the present over the annual cycle.

The Driving Forces of Climate

The source of energy that drives the climate is solar radiation (Figure 3). The Sun's energy travels across space as electromagnetic radiation to the Earth and determines the energy available for climate. Infrared (or "thermal") radiation, radio waves, visible light, and ultraviolet rays are all forms of electromagnetic radiation.

The Earth's atmosphere interferes with the incoming solar radiation (Figure 4, see page 7). About 31% of the radiation is reflected away by the atmosphere itself, by clouds, and by the surface. (The fraction of solar radiation a planet reflects back into space, and that therefore does not contribute to the planet's warming, is called its albedo. So the albedo of the Earth is about 31%.) Another 20% is absorbed by the atmosphere and clouds, leaving 49% to be absorbed by the Earth's surface.

To balance the incoming energy, the planet and its atmosphere must radiate, on average, the same amount of energy back to space (Figure 4). It does this by emitting infrared radiation. If the balance is upset in any way, for example, by a change in solar radiation, then

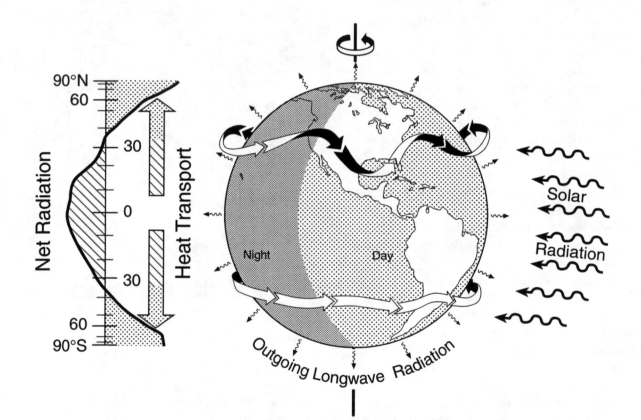

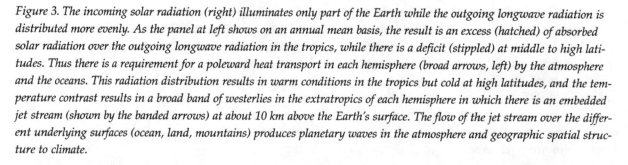

Figure 3. The incoming solar radiation (right) illuminates only part of the Earth while the outgoing longwave radiation is distributed more evenly. As the panel at left shows on an annual mean basis, the result is an excess (hatched) of absorbed solar radiation over the outgoing longwave radiation in the tropics, while there is a deficit (stippled) at middle to high latitudes. Thus there is a requirement for a poleward heat transport in each hemisphere (broad arrows, left) by the atmosphere and the oceans. This radiation distribution results in warm conditions in the tropics but cold at high latitudes, and the temperature contrast results in a broad band of westerlies in the extratropics of each hemisphere in which there is an embedded jet stream (shown by the banded arrows) at about 10 km above the Earth's surface. The flow of the jet stream over the different underlying surfaces (ocean, land, mountains) produces planetary waves in the atmosphere and geographic spatial structure to climate.

the Earth either warms or cools until a new balance is achieved. (Solar radiation, the electromagnetic spectrum, and the entire process of energy transfer between Sun and Earth are discussed in greater detail in the GCIP module *The Sun-Earth System*.) Most of the radiation emitted from the Earth's surface does not escape immediately into space because of the presence of the atmosphere and, in particular, because of the greenhouse gases and clouds in the atmosphere that absorb and re-emit infrared radiation.

Clouds play a complicated role in the planet's energy balance. They absorb and emit thermal radiation and have a blanketing effect similar to that of the greenhouse gases. They also reflect incoming sunlight back to space and thus

act to cool the surface. While the two opposing effects almost cancel each other out, the net global effect of clouds in our current climate, as determined by space-based measurements, is to cool the surface slightly relative to what would occur in the absence of clouds. Consequently, the bulk of the radiation that escapes to space is emitted either from the tops of clouds or by the greenhouse gases, not from the Earth's surface.

The Spatial Structure of Climate

Some parts of the Earth's surface receive more radiation than others (Figure 3). The tropics get the most, and actually gain more

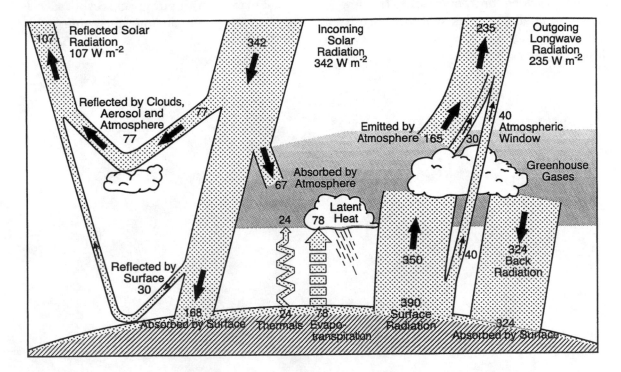

Figure 4. The Earth's radiation balance. The net incoming solar radiation of 342 W/m² (top center) is partially reflected by clouds and the atmosphere or by the Earth's surface (a total of 107 W/m², shown on the left-hand side of the figure). Of the remainder, 168 W/m² (49%) is absorbed by the surface. Some of that heat is returned to the atmosphere as sensible heating (indicated by thermals, bottom center) and some as evapotranspiration that is realized as latent heat in precipitation. The rest is radiated as thermal infrared radiation, and most of that is absorbed by the atmosphere and reemitted both up and down, producing the greenhouse effect (bottom right). The radiation lost to space comes from three sources. Some of it is emitted directly from the surface at certain wavelengths (40 W/m²); this region of the electromagnetic spectrum is called the "atmospheric window." Additional radiation is reflected to space from cloud tops (30 W/m²). The largest fraction (165 W/m²) comes from parts of the atmosphere that are much colder than the Earth's surface.

energy than they lose to space. The midlatitudes get less. The poles receive the least of all, emitting more energy than they receive from the Sun. This imbalance sets up an equator-to-pole temperature difference or "gradient" that results, when coupled with the influence of the Earth's rotation, in a broad band of westerly winds in each hemisphere in the lower part of the atmosphere. Embedded within these prevailing westerlies are the large-scale weather systems and winds from all directions (see Figure 6). These in turn, along with the ocean, act to transport heat poleward to offset the radiation imbalance (Figure 3). These weather systems are the familiar events that we see every day on television weather forecasts: eastward-migrating cyclones and anticyclones (i.e., low- and high-pressure systems) and their associated cold and warm fronts. Because they carry warm air toward the poles and cool air toward the equator, they are recognized as a vital part of the weather machine.

The continental land-ocean differences and obstacles such as mountain ranges also play a role by creating geographically anchored planetary-scale waves in the westerlies (Figure 3). These are the reasons why climate varies from, for instance, the west coast of the United States to the east coast. These waves are only semipermanent features of the climate system: they are evident in average conditions in any given year, but may vary considerably in their locations and general character from year to year. Specifically, changes in heating patterns can alter these waves and cause substantial regions of both above- and below-average temperatures in different places during any given season, such as the example given earlier for the winter of 1993–94.

II
The Weather Machine

Weather phenomena such as sunshine, clouds of all sorts, precipitation (ranging from light drizzle to rain to hail and snow), fog, lightning, wind, humidity, and hot and cold conditions can all be part of much-larger-scale weather systems. The weather systems are cyclones (low-pressure systems) and anticyclones (high-pressure systems) and the associated warm and cold fronts. Figure 5 gives a satellite image of a major storm system on the east coast of the United States. Accompanying panels show the temperatures that delineate the cold front (see below) and the sea-level pressure contours. It is systems like these, and their associated weather phenomena, that make up the weather machine.

Weather systems exist in a broad band both separating and linking warm tropical and subtropical air and cold polar air. They not only divide these regions but also act as an efficient mechanism for carrying warmer air toward the poles and cold air toward the equator. Thus, in the Northern Hemisphere, southerlies (winds from the south) are typically warm and northerlies (winds from the north) are cold. Within a weather system, the boundary of a region where warm tropical or subtropical air advances poleward is necessarily a region of strong temperature contrast. This boundary is called a warm front. As the warm air pushes cooler air aside, it tends also to rise, because warm air is less dense. Because the rising air also moves to regions of lower pressure it expands and cools, so that moisture condenses and produces clouds and rain.

The advancement equatorward of cold air occurs similarly along a cold front, but in this case, the colder and therefore denser air pushes under the somewhat warmer air in its path, forcing it to rise, often causing convective

clouds, such as thunderstorm clouds, to form. Note that the movement poleward of warm air and the movement equatorward of cold air usually go together as part of the same system because otherwise air would pile up in some places, leaving holes elsewhere.

The process of warm air rising and cold air sinking is pervasive in the atmosphere and is also a vital part of the weather machine. Warm air is less dense than cold air and is thus naturally buoyant. As seen in Figure 4, warmth is generally transferred from the surface to higher levels in the atmosphere, where the heat is eventually radiated to space. The process of transferring heat upward is called convection. It gives rise to a vast array of weather phenomena, depending on the geographic location, the time of year, and the weather system in which the phenomena are embedded. Clouds that result from convection are called convective clouds. These range from small puffy cumulus clouds, to multicelled cumulus that produce rain showers, up to large cumulonimbus clouds that may produce severe thunderstorms.

Weather systems over the oceans have a somewhat different character from those over land because of the abundant moisture over the oceans which more readily allows clouds and rain to form. Over land, storms are often more violent, in part because the land can heat and cool much more rapidly than the ocean and also because mountain ranges can create strong winds and wind direction changes (called wind shear) that can help facilitate the development of intense thunderstorms and even tornadoes. These conditions often occur in the United States in spring to the east of the Rocky Mountains, where northward-moving air has an

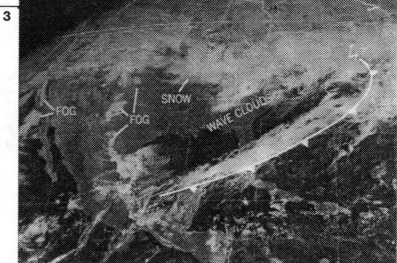

Figure 5. Satellite imagery of a major storm system on the East Coast of the United States (Panel 3). Panel 1 shows the temperatures that delineate the cold front, and Panel 2 gives the sea-level pressure contours in millibars. In Panel 1, cold air over the United States is pushing south and east, carried by strong northwesterly winds. Panel 2 shows the low-pressure cyclone system over the East Coast, which has a cold front attached, indicating the leading edge of the cold air. High pressures and an anticyclone exist over the northern Great Plains, accompanied by clear skies (Panel 3). The cloud associated with the cold front is also shown in Panel 3, along with many other weather phenomena typical in such cases, as marked on the figure. From Gedzelman (1980).

abundant supply of moisture (a prerequisite for cloud development) from the Gulf of Mexico. Weather phenomena and the larger weather systems develop, evolve, mature, and decay largely as turbulent instabilities in the flow of the atmosphere. Some of these instabilities arise from the equator-to-pole (i.e., horizontal) temperature contrast (Figure 6). If, for some reason, the contrast becomes too large, the situation becomes unstable, and any disturbance can set off the development of a weather system. Other types of instability occur as a result of vertical temperature gradients—often associated with warm air rising and cold air sinking (convective instability). These types of instability may be related to the warming of the surface air from below, or the pushing of warm and cold air masses against one another as part of a weather system developing. They may also occur as part of the cycle of night and day. Many other weather phenomena arise from other instabilities or from breezes set up by interactions of the atmosphere with complex surface topography.

Weather phenomena and weather systems mostly arise from tiny initial perturbations that grow into major events. The atmosphere, like any other system, is averse to unstable situations. This is why many triggering mechanisms exist that will push the atmosphere back toward a more stable state in which temperature contrasts are removed. In general, therefore, once the atmosphere has become unstable, some form of atmospheric turbulence will take place and grow to alleviate the unstable state by mixing up the atmosphere. It is not always possible to say which initial disturbance in the atmosphere will grow, only that one will grow. There is, therefore, a large component of unpredictable behavior in the atmosphere, an unpredictability that is exacerbated by and related to the underlying random component of atmospheric motions. The processes giving rise to this randomness are now referred to in mathematics as chaos. Because of the above factors, weather cannot be accurately forecast beyond about ten days.

The processes and interactions in the atmosphere are also very involved and complicated.

This aspect of atmospheric behavior is referred to as "nonlinear," meaning that the relationships are not strictly proportional. They cannot be charted by straight lines on a graph. The relationships in nonlinear systems change in disproportionate (and sometimes unpredictable) ways in response to a simple change. A gust of wind may be part of a developing cloud that is

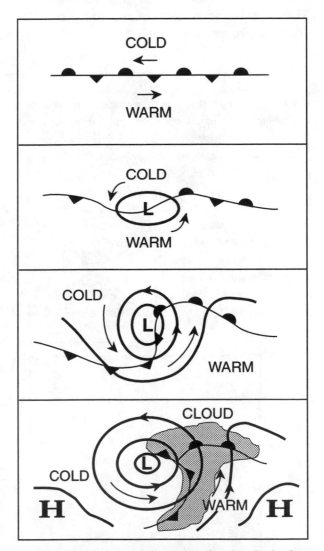

Figure 6. Baroclinic instability is manifested as the development of a storm from a small perturbation in the Northern Hemisphere with associated cold fronts (triangles) and warm fronts (semicircles). The arrows indicate the direction of wind. The shading on the bottom panel indicates the extensive cloud cover and rain or snow region in the mature stage.

embedded in a big thunderstorm as part of a cold front, which is attached to a low-pressure system that is carried along by the overall westerly winds and the jet stream (an example is given in Figure 5). All these phenomena interact and their evolution depends somewhat on just how the other features evolve.

Nevertheless, on average, we know that weather systems must behave in certain ways. There are distinct patterns related to the climate. So, while we may not be able to predict the exact timing, location, and intensity of a single weather event more than ten days in advance, because they are a part of the weather machine, we should be able to predict the average statistics, which we consider to be the climate. The statistics include not only averages but also measures of variability and sequences as well as covariability (the way several factors vary together). These aspects are important, for instance, for water resources, as described in Weather Sequences (see page 22).

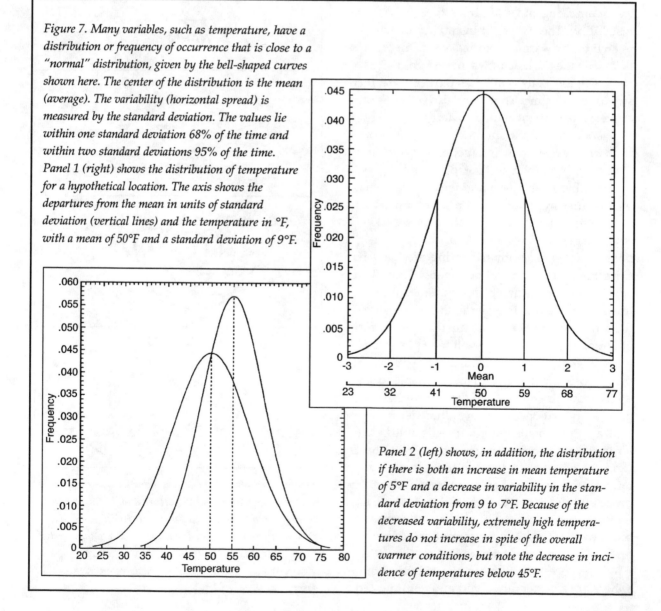

Figure 7. Many variables, such as temperature, have a distribution or frequency of occurrence that is close to a "normal" distribution, given by the bell-shaped curves shown here. The center of the distribution is the mean (average). The variability (horizontal spread) is measured by the standard deviation. The values lie within one standard deviation 68% of the time and within two standard deviations 95% of the time. Panel 1 (right) shows the distribution of temperature for a hypothetical location. The axis shows the departures from the mean in units of standard deviation (vertical lines) and the temperature in °F, with a mean of 50°F and a standard deviation of 9°F.

Panel 2 (left) shows, in addition, the distribution if there is both an increase in mean temperature of 5°F and a decrease in variability in the standard deviation from 9 to 7°F. Because of the decreased variability, extremely high temperatures do not increase in spite of the overall warmer conditions, but note the decrease in incidence of temperatures below 45°F.

To consider a more concrete example, suppose that the average temperature in a month is 50°F. In addition to this fact, it is also useful to know that the standard deviation of daily values is 9°F (Figure 7). This is the statistician's way of saying that 68% of the time the temperatures fall within 50 plus or minus 9, or between 41° and 59°F, and 95% of the time the values are expected to fall within 50 plus or minus 18, or between 32° and 68°F. We may also wish to know that the lowest value recorded in that month is 22°F and the highest 79°F. Moreover, if the temperature is above 60°F one day, we can quantify the likelihood that it will also be above 60°F the next day. And so on.

III
Climate Change

We have shown how climate and weather are intimately linked and explained how climate may be considered as the average of weather together with information about its variability and extremes. Climate, however, may be forced to change, not through internal weather effects, but due to the influence of external factors. And, if the climate changes in this way, so too will its underlying statistical nature, as characterized by the weather we experience from day to day. We now address this possibility.

Human-Caused Climate Change

The climate can shift because of natural changes either within the climate system (such as in the oceans or atmosphere) or outside of it (such as in the amount of solar energy reaching the Earth). Volcanic activity is an Earth-based event that is considered outside of the climate system but that can have a pronounced effect on it.

An additional emerging factor is the effect of human activities on climate. Many of these activities are producing effects comparable to the natural forces that influence the climate. Changes in land use through activities such as deforestation, the building of cities, the storage and use of water, and the use of energy are all important factors locally. The urban heat island is an example of very local climate change. In urban areas, the so-called concrete jungle of buildings and streets stores up heat from the Sun during the day and slowly releases it at night, making the nighttime warmer (by several degrees F in major cities) than in neighboring rural regions. Appliances, lights, air conditioners, and furnaces all generate heat. Rainfall on buildings and roads quickly runs off into gutters and drains, and so the ground is not moist, as it would be if it were an open field. By contrast, when the Sun shines on a farmer's field, heat usually goes into evaporating surface moisture rather than increasing the temperature; the presence of water acts as an air conditioner. In fact, in some places a reverse of urban warming, a suburban cooling effect, has been found because of lawns and golf courses that are excessively watered. Changes in the properties of the surface because of changes in land use give rise to these aforementioned climate changes. Nevertheless, these effects are mostly rather limited in the areas they influence.

The Enhanced Greenhouse Effect

Of most concern globally is the gradually changing composition of the atmosphere caused by human activities, particularly changes arising from the burning of fossil fuels and deforestation. These lead to a gradual buildup of several greenhouse gases in the atmosphere, with carbon dioxide being the most significant. They also produce small airborne particulates—aerosols—that pollute the air and interfere with radiation. Because of the relentless increases in several greenhouse gases, significant climate change will occur—sooner or later. The greenhouse-gas component of this change in climate is called the enhanced greenhouse effect. While this effect has already been substantial, it is extremely difficult to identify in the past record. This is because of the large natural variability in the climate system, which is large enough to have appreciably masked the slow human-produced climate change.

The amount of carbon dioxide in the atmosphere has increased by more than 30% (Figure 8) since the beginning of the industrial revolution, due to industry and the removal of forests. In the absence of controlling factors, projections are that concentrations will double from pre-industrial values within the next 60 to 100 years. Carbon dioxide is not the only greenhouse gas whose concentrations are observed to be increasing in the atmosphere from human activities. The most important other gases are methane, nitrous oxide, and the chlorofluoro-carbons (CFCs).

Effects of Aerosols

Human activities also put other pollution into the atmosphere and affect the amount of aerosols, which, in turn, influences climate in several ways. From a climate viewpoint, the most important aerosols are extremely small: in the range of one ten-millionth to one millionth of a meter in diameter. The larger particles (e.g., dust) quickly fall back to the surface.

Aerosols reflect some solar radiation back to space, which tends to cool the Earth's surface. They can also directly absorb solar radiation, leading to local heating of the atmosphere and, to a lesser extent, contributing to an enhanced greenhouse effect. Some can act as nuclei on which cloud droplets condense. Their presence therefore tends to affect the number and size of droplets in a cloud and hence alters the reflection and absorption of solar radiation by the cloud.

Aerosols occur in the atmosphere from natural causes; for instance, they are blown off the surface of deserts or dry regions. The eruption of Mt. Pinatubo in the Philippines in June 1991 added considerable amounts of aerosol to the stratosphere, which scattered solar radiation, leading to a global cooling for about two years. Human activities that produce aerosols include biomass burning and the operation of power

plants. The latter inject sulfur dioxide into the atmosphere, a molecule that is oxidized to form tiny droplets of sulfuric acid. In terms of their climate impact, these sulfate aerosols are thought to be extremely important; they form the pervasive milky haze often seen from air-craft windows as one travels across North America. Because aerosols are readily washed out of the atmosphere by rain, their lifetimes are short—typically a few days up to a week or so. Thus, human-produced aerosols tend to be concentrated near industrial regions.

Aerosols can help offset, at least temporarily, global warming arising from the increased greenhouse gases. However, their influence is regional and they do not cancel the global-scale effects of the much longer-lived greenhouse gases. Significant climate changes can still be present.

The Climate Response and Feedbacks

Some climate changes intensify the initial effect of greenhouse gases and some diminish it. These are called, respectively, positive and negative feedbacks, and they complicate the way the climate responds. For example, water vapor is a

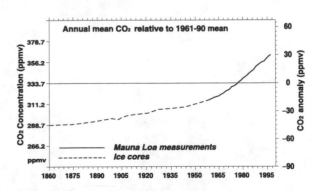

Figure 8. Annual carbon dioxide concentrations in parts per million by volume (ppmv). The total values are given at left, and the departures from the 1961–90 average (called anomalies) are given at right. The solid line is from measurements at Mauna Loa, Hawaii, and the dashed line is from bubbles of air in ice cores.

powerful greenhouse gas and therefore absorbs infrared radiation, so when a warmer climate causes more moisture to evaporate, the resulting water vapor increase will make the temperature even warmer. Clouds can either warm or cool the atmosphere, depending on their height, type, and geographic location. Hence they may contribute either positive or negative feedback effects regionally; their net global effect in a warmer climate is quite uncertain as it is not clear just how clouds may change with changing climate. Other important feedbacks occur through atmospheric interactions with snow and ice, the oceans, and the biosphere. Quantifying these various feedbacks is perhaps the greatest challenge in climate science, and the uncertainties in their magnitude are the primary source of uncertainty in attempts to predict the large-scale effects of future human-induced climate change.

IV
Observed Weather and Climate Change

Observed Climate Variations

Scientists expect climate change, but what changes have they observed? Analysis of global observations of surface temperature show that there has been a warming of about 0.6°C over the past hundred years (Figure 9). The trend is

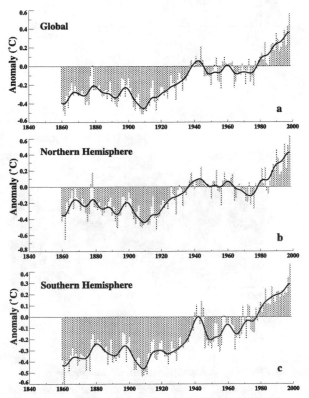

Figure 9. Average annual mean temperatures, expressed as anomalies from the 1961–90 average, over the Northern and Southern Hemispheres (middle and bottom panels) and for the globe from 1860 to 1998. Mean temperatures for 1961–90 are 14°C for the globe, 14.6°C for the Northern Hemisphere, and 13.4°C for the Southern Hemisphere. Based on Jones et al. (1999).

toward a larger increase in minimum than in maximum daily temperatures. The reason for this difference is apparently linked to associated increases in low cloudiness and to aerosol effects as well as the enhanced greenhouse effect. Changes in precipitation and other components of the hydrological cycle are determined more by changes in the weather systems and their tracks than by changes in temperature. Because such weather systems are so variable in both space and time, patterns of change in precipitation are much more complicated than patterns of temperature change. Precipitation has increased over land in the high latitudes of the Northern Hemisphere, especially during the cold season.

Figure 10 shows changes observed in the United States over the past century. Note especially the trend for wetter conditions after about the mid-1970s in the first panel (a). Panel b reveals that the main times of drought in the United States were in the 1930s and the 1950s. In the 1930s there was extensive drying in the Great Plains, referred to as the Dust Bowl because of the blowing dust and dust storms characteristic of that time. In part, the Dust Bowl was exacerbated by poor farming practices.

Naturally, times of moisture surplus tend to alternate with times of extensive drought. Panel c reveals the increasing tendency for rainfall to occur in extreme events of more than two inches of rain per day over more of the country. Thus, heavy rainfalls tend to occur more often or over more regions than previously, a steady and significant trend of about a 10% increase in such events. Temperatures have also increased in general (Panel d), but the warmest years tend to be those associated with the big droughts, which

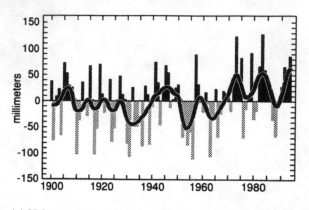

(a) U.S. average annual precipitation

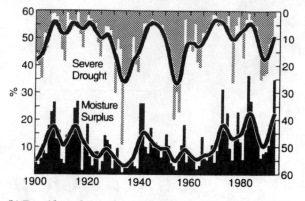

(b) Dry (drought) and wet (flood) conditions in the U.S.

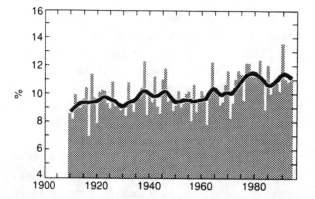

(c) Percent U.S. much above normal rainfall from 1 day extreme events (>2")

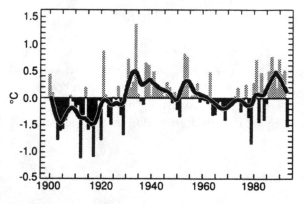

(d) U.S. Temperatures

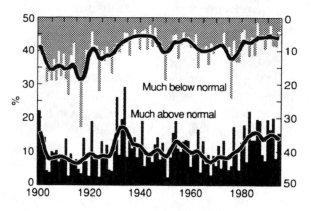

(e) Much above and much below normal U.S. temperatures

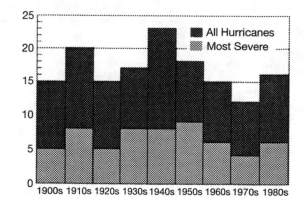

(f) Number of hurricanes making landfall

Figure 10. (a) The variations in U.S. average annual precipitation from the long-term average (mm), (b) the incidence of droughts and floods expressed as percentages of the U.S. land area, (c) the percentage of the United States that receives more than 2 inches (50.8 mm) of rainfall in one day, (d) the variation of the average annual U.S. temperature from the long-term average (°C), (e) the incidence of much-above and much-below normal temperatures expressed as percentage of U.S. land area, and (f) the number of hurricanes making landfall. In panels b, c, and e, the definitions of drought, flood, much above normal, and much below normal all correspond to the top or bottom 10% of all values on average. In panels b and e, the extents of the much-above and much-below normal areas are plotted opposite one another as they tend to vary inversely. This is not guaranteed, however, as wet (warm) conditions in one part of the country can be and often are experienced at the same time as dry (cold) conditions elsewhere (see Figure 11). From Karl et al. (1995).

contribute to many heat waves (because water is no longer present to act as a natural air conditioner). In the United States, some of the warmest years occurred in the 1930s. The warmth of the 1980s and 1990s, especially compared with the 1900 to 1920 period, cannot simply be explained by heat waves and changes in drought, however. In Panel e, we see that most of the temperatures much below average occurred in the early part of this century, while most of the temperatures well above average occurred either in the past 15 years or in the 1930s. Hurricanes naturally vary considerably in number from year to year (Panel f). Since they are so variable, and relatively rare, no clear trends emerge.

Figure 11 shows a consolidation of these factors into a U.S. climatic extremes index (CEI). It is made up of the annual average of whether several indicators are much-above or much-below normal, where these categories correspond to the top and bottom 10% of values. A value of 0% for the index would mean no portion of the country experienced extreme conditions in any category. A value of 100% would mean the entire country was under extreme conditions throughout the year under all categories. The average value, because of the way the index is defined, must be around 10%, and

the variations about this value indicate the extent to which the country was experiencing an unusual number of extremes of one sort or another. The major droughts of the 1930s and 1950s again are evident in this figure. In more recent decades, the increase in extremes comes from the increases in much-above normal temperatures and the increase in extreme one-day rainfall events exceeding two inches.

Interannual Variability

A major source of variability from one year to the next is El Niño. The term El Niño (Spanish for the Christ child) was originally used along the coasts of Ecuador and Peru to refer to a warm ocean current that typically appears around Christmas and lasts for several months. Fish yields are closely related to these currents, which determine the availability of nutrients, so the fishing industry is particularly sensitive to them. Over the years, the term has come to be reserved for those exceptionally strong warm intervals that not only disrupt the fishing industry but also bring heavy rains.

El Niño events are associated with much larger-scale changes across most of the Pacific

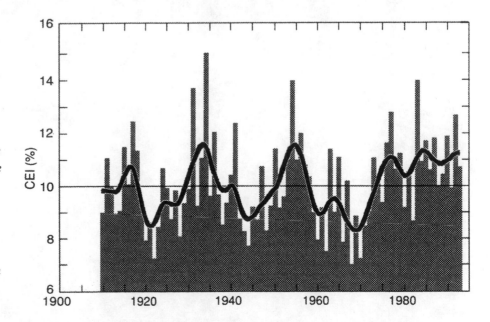

Figure 11. The CEI is the sum of two numbers. The first reflects the percentage of the United States, by area, where maximum and minimum temperatures, moisture, and days of precipitation were much-above or much-below normal. The second number is twice the percentage of the United States, by area, where the number of days of very heavy precipitation (more than two inches) was much greater than normal. From Karl et al. (1995).

Ocean. These changes in turn alter weather patterns around the globe through changes in the atmospheric circulation. They can alter the atmospheric waves (Figure 3) and thus the tracks of storms across North America and elsewhere. The major floods in the summer of 1993 in the upper Mississippi River basin were partly caused by El Niño. Recent floods in California (winters of 1994–95 and 1997–98) were also linked to El Niño as the storm track continually brought weather systems onto the west coast of the United States. (El Niño is discussed more extensively in the module *El Niño and the Peruvian Anchovy Fishery*.)

Because the magnitude of El Niño events is relatively large compared with climate change on the slower decadal time scale, El Niño is manifested much more readily than global warming in the weather we experience and in the regional climate variations. This is a prime example of interannual variability of climate, which, in general, tends to mask the climate change associated with global warming.

V
Prediction and Modeling of Climate Changes

In general, climate changes cannot be predicted simply by using observations and statistics. They are too complex or go well beyond conditions ever experienced before. For the most detailed and complicated projections, scientists use computer models of the climate system called numerical models. These models are based on physical principles, expressed as mathematical formulas and evaluated using computers.

Climate Models

Global climate models attempt to include the atmospheric circulation, oceanic circulation, land surface processes, sea ice, and all other processes indicated in Figure 1. They divide the globe into three-dimensional grids and perform calculations to represent what is typical within each grid cell. For climate models, owing to limitations in today's computers, these grid cells are quite large—typically 250 kilometers in the horizontal dimension and a kilometer in the vertical dimension. As a result, many physical processes can only be crudely represented by their average effects.

One method used to predict climate is to first run a model for several simulated decades without perturbations to the system. The quality of the simulation can then be assessed by comparing the average, the annual cycle, and the variability statistics on different time scales with observations. If the model seems realistic enough, it can then be run including perturbations such as an increase in greenhouse-gas concentrations. The differences between the climate statistics in the two simulations provide an esti-

mate of the accompanying climate change.

To make a true prediction of future climate it is necessary to include all the human and natural influences known to affect climate (cf. Figures 1 and 12). Because future changes in several external factors, such as solar activity and volcanism, are not known, these must be assumed to be constant until such time as we are able to predict their changes.

Climate Predictions

The climate is expected to change because of the increases in greenhouse gases and aerosols, but exactly how it will change depends a lot on our assumptions concerning future human actions. When developing countries industrialize, they burn more fossil fuels, generate more electricity, and create industries, most of which produce some form of pollution. Developed countries are currently the largest sources of pollution and greenhouse gases. Because future changes are not certain, climate models are used to depict various possible "scenarios." These are not really predictions but projections of what could happen. If a projection indicates that very adverse conditions could happen, policy actions could be taken to try to change the outcome. The following are some features of possible future climate changes created by human activities. Greatest confidence exists on global scales; regional climate changes are more uncertain.

1. The models indicate warming of 1.5 to 4.5°C for a climate with atmospheric CO_2 concentrations doubled from preindustrial times, when they were 280 parts per million by vol-

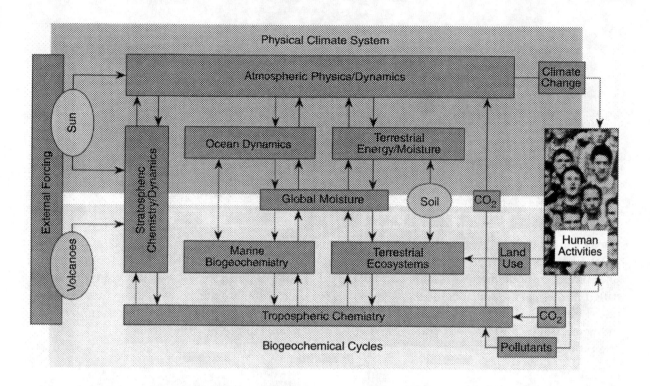

Figure 12. Schematic model of the fluid and biological Earth that shows global change on a time scale of decades to centuries. A notable feature is the presence of human activity as a major inducer of change; humanity must also live with the results of change from both anthropogenic and natural factors. From Trenberth (1992).

ume. An effective doubling of CO_2, taking into account aerosols and other greenhouse gases, is likely to occur around the middle of the 21st century. Corresponding manifestations of Northern Hemisphere climate change will take place some 20 to 50 years later because it takes the oceans at least that long to respond. The lag is likely to be greater over the Southern Hemisphere because of the influence of the larger ocean area. Aerosols are also expected to increase in areas undergoing industrialization (such as China) and to decrease in North America and Europe, where steps are being taken to decrease acid rain by decreasing sulfur emissions. The effects of aerosols will complicate climate change and will most likely change the regional distribution of the temperature increase. When effects of aerosols and greenhouse gases are combined, one estimate puts the average rate of temperature increase in the next century at about 0.15 to 0.25°C per decade. Such a warming

is expected to lead to an increase in extremely hot days and a decrease in extremely cold days.

So far, over the past century, during which time carbon dioxide has increased from 290–300 to 360 parts per million by volume (roughly a 20% increase), the observed temperature increase has been fairly modest, about 0.5°C (see Figure 9). This temperature increase is reasonably consistent with model predictions when effects of aerosols are included. But large uncertainties remain, particularly because of questions about how clouds might change.

2. The hydrological cycle is likely to speed up by about 10% with CO_2 doubling, bringing increased evaporation and increased rainfall in general. With warming, more precipitation is apt to fall as rain in winter instead of snow, and, with faster snowmelt in spring, there is likely to be less soil moisture at the onset of summer over midlatitude continents. When this change is combined with increased evaporation in sum-

mer, any natural tendency for a drought to occur is likely to be enhanced. However, there is not good agreement among the models on this aspect. An enhanced hydrological cycle also implies increased intensity of rainfall, such as has been found for the United States (Figure 10). Increases in rainfall in winter but drier conditions in summer would challenge future water managers to avoid flood damage and keep up with the demand for fresh water.

3. Because warming causes the ocean to expand and snow and glacial ice to melt, one real threat is a rise in sea level. There may be some compensation through increased snowfall on top of the major ice sheets (Greenland and Antarctica) so that they could increase in height even as they melt around the edges. Currently, sea level is observed to be rising by 1 to 2 mm/year, and this rate should increase, so that there are prospects for about a 50-cm rise in sea level by 2100, but the main impacts are not likely to be felt until the 22nd century.

4. With warming, increases in water vapor (a greenhouse gas) and decreases in snow cover and sea ice (lower albedo) provide positive feedbacks that should enhance the warming as time goes on. The land should warm more than the oceans, and the largest warming should occur in the Arctic in winter.

5. Stratospheric cooling is another likely effect of increased greenhouse gases. This cooling has important implications for ozone depletion, because the chemistry responsible for the Antarctic ozone hole is more effective at lower temperatures. The loss of ozone also increases stratospheric cooling.

6. Because of increased sea-surface temperatures, there may be changes in tropical storms and hurricanes. Hurricanes sustain themselves at temperatures above 27°C, feeding on the extra water vapor and latent heat those temperatures create. However, natural variability of hurricanes is large (Figure 10), so any effect from climate change will be hard to detect for many decades.

7. Coupled ocean-atmosphere general circulation models have only very recently been able to simulate rudimentary El Niño cycles. It seems likely that El Niño will continue to exist in a warmer world. Because El Niño and its cool counterpart La Niña create droughts and floods in different parts of the world, and because global warming tends to enhance the hydrological cycle, there is a real prospect that future such events will be accompanied by more severe droughts and floods. In the tropics, in particular, because of the great dependence on thunderstorm rainfall and its tendency to fall at certain times of the year (during the wet or monsoon season) the main prospect that looms is one of larger variability and larger extremes in weather events.

Interpretation of Climate Change in Terms of Weather

For assessing impacts, what is most needed are projections of local climate change. However, producing such projections represents a considerable challenge. Climate predictions are especially difficult regionally because of the large inherent natural variability on regional scales. We have discussed changes in climate mostly in terms of changes in average conditions. But we experience those changes mainly through changes in the frequency of extreme weather events, e.g., how hot it gets on a daily basis, or how frequent and violent thunderstorms become. An average monthly change in temperature of 3°C (5°F) may not sound like very much, but it has a very dramatic effect on the daily frequency of extreme temperatures, e.g., see Figure 7. For example, currently in Des Moines, Iowa, the likelihood that the maximum temperature on any day in July will exceed 35°C (95°F) is about 11%. However, with an increase in the average monthly maximum temperature of 3°C the likelihood almost triples, to about 30%. Small changes in the average can bring about relatively large changes in frequencies of extremes.

In addition to a change in the average climate, the variability itself could also change. If the daily variability of temperature increases in

Des Moines, then an even greater portion of days would exceed 35°C. If, on the other hand, the variability decreases, the temperature from one day to another would be more similar than before. Changes in variability affect changes in the frequency of extremes and have more effect than changes in averages (see Figure 7). There is some evidence that with climate warming, daily variability of temperature might decrease so that there might be fewer cold extremes in winter. Variability of temperature could decrease in some seasons (e.g., winter) but increase in others.

Changes in variability of precipitation are also anticipated and will tend to be associated with changes in the average precipitation. Variability generally increases as average precipitation increases. In the United States, precipitation extremes have been found to increase in the past few decades (see Figure 10 Panel c). The picture for precipitation is more complicated, however. For example, climate change is likely to alter the jet stream and associated location of storm tracks, so that some places will experience an increase in storminess while others, not very far away, will experience a decrease. Such opposite changes over short distances should be expected and are an inherent part of climate for rainfall, but this is likely to be confusing to many people.

It is likely that most people in developed countries will continue to experience weather much as they have before. In some places they may notice that the time between major snow storms is longer, heat waves are more frequent and debilitating, the intensity and frequency of thunderstorms are changed, coastal damage to beaches is more common, prices of some commodities increase while others decrease, water conserving practices in certain communities are intensified, and so on. Areas where the cumulative effects of weather are important, such as water resources and agriculture, may be more at risk.

Many of the effects may be rather subtle most of the time, and the actual impact may originate through other pressures (increasing population, as an example) and may only be exacerbated by the changes in climate. But there are also likely to be dramatic effects. As an example, during a drought a string of widespread heat waves may put increased demand on air conditioning, causing brownouts and even blackouts as the electricity demand exceeds available capacity; or there may be more medical emergencies, such as heat stroke, involving those who do not have or cannot afford air conditioning. Ironically, the extra use of air conditioning leads to increased fossil fuel use and hence a greater emission of greenhouse gases.

VI
Impacts of Weather and Climate Changes on Human Activities

Human activities and many sectors of economic activity depend on weather and climate in different ways. Some rely on average conditions. Others are sensitive to extremes. Yet others depend upon variety and so weather sequences can be important. Aside from choosing the climate by selecting the right location, there are other ways we can attempt to cope with climate change and its consequences for agriculture, fisheries, and so forth.

Weather Sequences

Conditions may be altered not only by individual weather events but also by sequences of weather events. Weather sequences, for example, play a big role in determining stream runoff and soil moisture, and can result in prolonged periods of abnormal temperatures and sunshine. These are important determinants of agricultural yields, and the responsiveness of yields to such other inputs as fertilizer depends on the growing conditions supplied by a sequence of weather events.

Runoff to surface streams and groundwater recharge, or replenishment, depend on extended sequences of weather events so that the contribution of individual rainstorms to runoff depends on whether previous conditions were wet or dry. In addition, the timing of runoff in mountainous river basins is strongly dependent on snowpack accumulation and rate of melt. Mountain runoff, thus, is quite sensitive to temperature variations. The quantity and timing of runoff, in turn, determine the availability of water for competing agricultural, municipal, industrial, hydropower, recreational, and ecological uses.

As an example, suppose place A has 0.5 inches of gentle rain every three days, for a monthly average of 5 inches, and place B has 2.5 inches of rain on two consecutive days of the month but with all other days dry, again for a monthly total of 5 inches. The monthly total is the same, but the sequence differs greatly and the climates would be quite different. At place A, the rain would replace the evaporation and use of moisture by plants; there would be few puddles, so there would be no runoff into streams. As a rule of thumb, anytime there is more than 3 inches of rain in a day, there will be fairly extensive flooding. So at place B it is likely that low-lying parts of roads would be flooded, culverts would overflow, basements would flood, and there would be substantial damage from all the runoff during the two rainy days. But then the rest of the month, the ground would dry out and plants would become stressed and wilt unless they had very deep and extensive roots. The different sequences of weather make for very different impacts.

Location, Location, Location

Climate and weather contribute to personal satisfaction. For example, the satisfaction provided by a walk in the park varies according to whether conditions are balmy or blustery. A simple economic model of the allocation of time between walks in the park and other activities predicts that parks will become more crowded as the weather improves. Casual observations confirm that prediction. Many people also express a willingness to pay to live where they can expect to enjoy particular climatic characteristics, such as frequent mild, sunny weather. Their valuations of those characteristics may be expressed as a willingness to accept a somewhat lower real wage or to pay more for housing of comparable quality in order to live in a preferred climate.

Climates are tied to particular locations, so that when individuals decide to move themselves and their productive activities to a certain place, they are also choosing the climate in which they will live and operate. For most economic activities, climate is only one of many factors influencing choice of location. For some activities, the characteristics of climate are a central factor in location decisions. The expected availability of snow is an important concern for the location of ski resorts. A sufficiently low risk of severe freezes is a critical consideration in the location of orange groves, and crop selection decisions and farm management strategies are heavily influenced by probable growing-season conditions.

The location of other industries is tied to the availability of particular natural resources. The lumber and paper industries require trees. Hydropower dams are located where stream gradients and rates of flow offer significant potential generation. Fishing fleets and processing capacity are based to allow access to expected concentrations of commercially valuable fish. Such resources are themselves tied to climate. The connections are obvious for hydropower, where drought conditions can quickly lead to reduced generation. The impacts of climatic variations on the timber industry are less immediate, although prolonged droughts can significantly reduce the stock of healthy standing trees and often create favorable conditions for forest fires.

Severe Weather Events

The most dramatic impact of weather on human endeavors is often through severe weather events that may alter as the climate changes. Severe weather has always affected human activities and settlements as well as the physical environment. It can damage property, cause loss of life and population displacement, destroy or sharply reduce agricultural crop yields, and temporarily disrupt essential services such as transportation, telecommunications, and energy and water supplies. Society has developed various methods to avoid or minimize adverse impacts of weather and has also developed means to facilitate recovery from extreme weather phenomena. Yet, because severe weather events repeatedly disrupt socioeconomic activities and cause damage, society continues to search for new ways to protect lives and property. Some of these involve behavioral adjustments based on past societal experience, such as educating citizens about what to do in the event of a tornado warning. Others involve the application of new meteorological research findings for improving the prediction of where and when severe weather will occur (see page 28).

Societal Responses

One way of reducing vulnerability to weather is to reduce damage to property, through such strategies as stricter construction standards, tighter building codes, and restrictions on development in floodplains and on coastal barrier islands. The construction of storm sewers can help minimize short-term flood damage in highly developed areas where there is substantial impermeable surface such as pavement. The casualty and hazard insurance industry in more developed countries helps insured parties rebuild and replace property damaged by severe weather. Of course, insurance does not physically protect property from weather-related damage, but it does facilitate recovery and replacement in the aftermath of extreme weather events such as tornadoes, hurricanes, and floods. The insurance industry itself has been altered by perceptions of climate change, such as rates for coastal insurance in Florida. Reservoirs increase resilience to short-term fluctuations in streamflows and thus protect the water supply and hydropower production. Electric utilities also increase resilience to variable hydropower output and variable demand by maintaining backup generation capacity (e.g., coal-fired plants) and by buying or selling power over interconnected transmission grids.

A second way of reducing vulnerability to weather is through technology. Many technologies are so common that they have become part of society's everyday affairs and activities. For example, modern tires, windshield wipers, and fog lights have helped reduce the hazard of driving in bad weather conditions. Indoor heating and air conditioning provide comfort and protection from extreme temperatures in winter and summer. The invention of shelter itself was probably prompted by human desires to have protection from the extremes of weather and climate as well as from predators and human enemies.

Modern weather forecasting, which has progressed rapidly over the past half-century, can give advance warning of possibly dangerous weather conditions. Forecasters can frequently provide information minutes to several days ahead of possible severe weather conditions. In many cases, decisions may be made based on forecasts to reduce or eliminate potential vulnerability to severe weather. For example, on a construction site, concrete deliveries may be rescheduled to ensure that snow, ice, and cold temperatures do not interfere with its proper curing. Businesses may alter trucking schedules and routes in response to anticipated foul weather. In certain circumstances, farmers may be able to harvest all or part of their crops in advance of what could be destructive weather. The usefulness of weather forecast information varies among economic sectors. While a reliable weather forecast may help a farmer to efficiently schedule crop irrigation, for example, it cannot help that farmer protect a crop from imminent hail damage. Other coping mechanisms, such as crop insurance, preparedness, and routine maintenance of flood levees and storm sewers, also help society manage its vulnerability to extreme weather events.

Managing Risk

Climate and day-to-day weather variations affect a wide variety of economic activities. Climate influences the spatial distributions of population and of industrial, agricultural, and resource-based production activities, while weather can affect levels of production and production costs. In addition, severe weather can damage or destroy property.

In gambling, even the most astute players will occasionally lose. In economics, if climate-induced loss reveals new information on the nature of the climatic risk or on the vulnerability of affected activities, or if it alters people's perceptions of the risk, then they will readjust their risk-management strategies. If not, they will go back to the status quo. For example, towns that are hit by tornadoes are usually rebuilt in the same location because one hit does not signal any change in the long-term risk. A series of extreme events, on the other hand, may be taken as a signal that previously available information provided an inaccurate picture of the true risk, or that the climate has changed. In that situation, a town might not rebuild in the same location.

Impacts on Agriculture

Humans have been interested in understanding and predicting the effects of climate on crop production since the rise of agriculture, because food production is critical to human survival. A classic Biblical example is in Genesis, where Joseph interprets a dream of the Pharaoh's as a portent of seven coming years of good grain harvests followed by seven years of crop failure.

Crop yields are strongly affected by changes in technological inputs such as fertilizer, pesticides, irrigation, plant breeding, and management practices, but the major cause of year-to-year fluctuations in crop yield is weather fluctuations. Agricultural crops are mainly sensitive to fluctuations in temperature and precipitation, although solar radiation, wind, and humidity are also important. In general a crop grows best and produces maximum yield for some optimum value of the relevant climate variable; as conditions depart from the optimum, the plants suffer stress. The responsiveness of yields, and therefore the financial return, to such inputs as fertilizer and pesticides varies with weather conditions, so that it is prudent for farmers to make adjustments depending on the weather.

Effects of Temperature and Precipitation on Crop Yield. The temperature regime of a particular locale will affect the timing of planting and harvesting and the rate at which the crop develops. With adequate moisture, the potential growing season is largely determined by temperature; in temperate mid-latitude regions this generally extends from the last frost in the spring to the first frost in the fall. The rate at which plants develop and move through their

Pacific Salmon

In the case of private production and investment decisions, the climate-related risks fall largely on the parties making the decisions unless they have chosen to purchase some form of insurance, allowing the sharing of the risk with others. To the extent that the decision makers bear the risk, they have the incentive to engage in appropriate risk-management strategies and to make efficient use of available climate- and weather-related information.

Many climate-sensitive natural resources are managed as public property, and decisions regarding their use are made by government agencies, often with considerable input from the interested public. In such cases, the effects of climatic variability often complicate the already difficult task of balancing the conflicting demands of competing interests. Fisheries are sensitive to climatic variations, but the true impacts of climate are often complex and difficult to separate from the impacts of other factors (such as fishery management, overfishing, spawning habitat degradation, water diversions, building of dams, and pollution) influencing the survival, growth, and spatial distribution of fish populations.

The Pacific salmon fishery provides an example. Since the mid-1970s, warmer sea-

surface temperatures along the Pacific coast of North America and changes in near-shore currents associated with more frequent and persistent El Niño events appear to have contributed to remarkable increases in the productivity of Alaskan salmon stocks and to declining runs of some salmon spawning in Washington, Oregon, and California. In the early 1990s, these trends culminated in a series of record Alaskan salmon harvests and severe declines in once-thriving Coho and Chinook fisheries in Washington and Oregon. These fluctuations in northern and southern salmon stocks contributed to the breakdown of international cooperation under the Pacific Salmon Treaty. Under pressure from commercial, sport, and Indian fishing interests within their respective jurisdictions, British Columbia, Alaska, and the West Coast states were unable to come to a consensus over a fair and biologically sound division of the harvest for six years. The resulting inability to control Alaskan and Canadian exploitation of depleted stocks migrating to the southern spawning areas contributed to their further decline. Finally, in June 1999, the governments responded to the imperiled state of the stocks by implementing a new agreement that adjusts harvests to changes in abundance.

growth stages (crop phenology) is regulated by temperature. The thermal requirements of crops are often determined by adding up the temperatures over time and determining the total thermal units, often referred to as growing degree days, that are required to complete particular growth stages. Temperature also affects the rate of plant respiration and partially determines the plant's need for water by determining the evapotranspiration rate.

Growing plants can be damaged by temperature extremes that interfere with their metabolic processes and may be especially sensitive during particular stages of growth. For example, in growing corn, severe high temperature stress for a ten-day period during silking (a critical phenological stage when the numbers of kernels on the corn ear is determined) can result in complete crop failure.

Water is necessary for plant growth, and so precipitation is also extremely important. All important physiological processes such as photosynthesis, respiration, and grain formation require moisture. Drought is certainly the weather extreme that has been most studied in terms of its impact on agriculture. Crops are particularly sensitive to moisture stress during certain phenological stages. For example, moisture stress is especially harmful to corn, wheat, soybean, and sorghum during the periods of flowering, pollination, and grain-filling. Inadequate moisture causes reduced crop yield.

Agriculture and Climate Fluctuations of the Past. In the 20th century there have been various periods of drought in North America, but the most serious was the prolonged drought of the 1930s in the Great Plains of the United States and Canada. The extremely low precipitation and relatively high temperatures (Figure 10) resulted in drastic reductions in grain yields. Wheat yields in Saskatchewan province in Canada for the years 1933–37 were less than half the yields obtained in the 1920s. In the south-central United States (Oklahoma, Kansas, Colorado, Texas, and New Mexico) rainfalls about 100 mm below normal for these

years and poor farming practices combined to produce a lot of blowing dirt and many severe dust storms, creating the Dust Bowl.

Another example is the prolonged cool period during the 16th and 17th centuries in Europe, known as the Little Ice Age. In its coldest phase, the average annual temperature in England was approximately 1.5°C less than that of the 20th century, resulting in widespread and frequent crop failure because of the greatly reduced growing season and cold damage to crops. In the hill country of southeast Scotland between 1600 and 1700 the oat crop failed on average one year out of every three.

Short-lived extreme temperatures can also severely affect crops. In the Corn Belt of the United States in 1983, substantial losses occurred because of blistering hot temperatures in July, including a week when maximum temperatures remained above 35°C, when the corn was flowering. This example indicates that even now, when farming is technologically advanced, extreme weather can result in serious losses.

Future Climate Change and Agriculture. Although there are many uncertainties regarding how climate may change due to increased greenhouse gases, there are some likely changes that would affect agriculture in specific ways. Generally, increased temperatures would bring about longer potential growing seasons, which would allow for multiple cropping in some areas (i.e., raising more than one crop per season). Also, crops would reach maturity more quickly; however, this could result in declining yields, since the crop would have less time to form grain. Higher temperatures would increase the respiration rates of plants (the process by which plants break down organic substances), thus reducing the amount of biomass available for yield formation. More-frequent high temperatures could also result in crop damage by increasing evaporation and hence moisture stress and wilting in plants even if there were no changes in precipitation.

Increased greenhouse warming is likely to

Florida Citrus

An instructive example is provided by the experience of Florida citrus growers with a series of devastating freezes during the 1980s. At the beginning of the 1980s, Florida's orange groves were concentrated in the center of the state, along the north-south central ridge. Near the northern edge of the citrus belt, freeze damage to fruit and occasional loss of trees were a common occurrence, but growers had learned to manage those risks: by balancing their investments in orange groves against other sources of income, by avoiding planting in known cold pockets, and by engaging in mitigative actions, such as turning on sprinklers as freezing temperatures approached.

Then January 1981 brought the first in a series of five tree-killing freezes that profoundly altered the central Florida landscape. The two most damaging freezes, in December 1983 and January 1985, together killed approximately one-third of the state's commercial citrus trees, virtually eliminating groves in several counties close to the former heart of the citrus belt. Lake County, which had the second-largest citrus acreage in the state at the beginning of 1980, lost more than 90% of its orange trees to the 1983 and 1985 freezes. The final freeze in the series, in December 1989, killed the majority of the trees that had managed to survive the earlier freezes as well as 61% of Lake County's newly replanted trees. Statewide, the 1989 freeze killed far fewer trees than did the 1983 and 1985 events, in part because of a major shift in new citrus planting to more southerly areas.

Why, you might ask, were the groves not located in those southern relatively freeze-safe counties to begin with? Because the heavy, wet soils there require expensive preparation and drainage before citrus can be planted. In addition, yields were traditionally lower, and trees were more prone to diseases in those areas. So growers had weighed the risk of freeze-related losses against the expected differences in net returns in their original decisions about where to locate groves. Their experiences during the 1980s increased their apparent wariness of the freeze risk, tilting the balance in favor of the southern growing areas, where millions of new citrus trees have been planted since the mid-1980s.

Such adjustments to long-term climatic variations and to new information regarding weather-related risks can be made more easily for some types of activities than for others. It is relatively easy to alter the mix of annual crops to be planted or the proportion of fallow to planted acreage if new information becomes available before the beginning of the planting cycle. A forecast of unusually hot, dry conditions over the growing season might induce farmers to leave a larger proportion of their land fallow and increase the proportion of the remainder devoted to drought-tolerant varieties.

Rapid adjustments are more difficult where the production process is not resilient to climatic variations and depends on relatively immobile capital assets. In the Florida citrus example, the trees were expected to be long-lived, immobile capital assets. Their destruction provided growers with the necessity and opportunity to rethink their investment strategies.

result in both increases and decreases in precipitation in different areas. Increased drought would bring about reduced yields in general and a greater likelihood of complete crop failure, particularly in areas of the world that are currently vulnerable to environmental changes and where water is already limited for agriculture, such as in the semiarid areas of Africa bordering the Sahara Desert.

The Effects of Nonclimatic Factors. Several factors could limit loss in agricultural productivity due to climate change. One is mitigation by the direct physiological effects of increased CO_2 on plants. If CO_2 were increasing without any change in climate, agricultural productivity worldwide would likely increase. Experiments indicate that some plants grown in atmospheres enriched in CO_2 show increased rates of photosynthesis and of net photosynthesis (total photosynthesis minus respiration). They tend to use water more efficiently and thus require less. These plants (the so-called C3 class), include wheat, rice, and soybeans. Other crops, such as corn and sorghum (in the C4 group), do not benefit from increased concentrations. If CO_2 increases, these species—which are particularly important in developing countries—may even be at a competitive disadvantage compared to weeds belonging to the C3 group.

Another factor is technological adaptations to environmental changes. Agriculture is an ecosystem managed by humans. Possible adaptations to climate change include increasing (or decreasing) irrigation, changing crop type to one more adapted to the new climate, breeding hybrids of the original crop that can better cope with the new climate (e.g., breeding for drought tolerance), and adjusting fertilizer, herbicide, and pesticide use. While many adaptations are possible, their success is difficult to determine, because the ultimate value of agricultural crops or changes in productivity can only be determined when considering the interactions of global economies. For example, let us say that with climate change, sorghum grows better in

the central Great Plains because it is more drought tolerant than wheat. Farmers may be able to switch crops, but the profitability depends on the demand for sorghum in domestic and international markets.

In recent years, climate change assessments of agriculture have become sophisticated enough to analyze the impact of climate change on agriculture, direct physiological effects, possible technological adaptations, and changes in the global economy. Such "integrated assessments" make numerous assumptions about the future, above and beyond how the climate may change. These studies are highly complex and rife with uncertainties. It seems that on a global basis developed countries may be able to adapt fairly well to climate change, but there is concern that some developing countries could suffer serious economic and human hardships.

Planning for Local Weather Changes

Society should develop appropriate responses to help manage and reduce vulnerability to extreme meteorological events. It is likely that increasing frequency and/or intensity of severe weather as a result of climate change will put more lives and property at risk, particularly in coastal and inland areas close to coastlines. Around the world, these areas have already become very vulnerable over the past several decades as human populations and development have grown dramatically in coastal and near-coastal regions. As an illustration, the population of U.S. coastal counties (Atlantic, Great Lakes, Gulf of Mexico, and Pacific) grew by nearly 32 million residents between 1960 and 1990 (from about 75 million to 107 million), according to the U.S. Bureau of the Census. Similar or even higher rates of coastal-zone population growth occurred in many other countries during the same period, making them more vulnerable to coastal storms and wave surges as well.

In response to more frequent and/or more intense weather events, societal rules governing such matters as building codes and land use could be modified or strengthened to help reduce vulnerability. Alterations in the insurance industry (e.g., changes in government regulation of the industry) could help restructure financial instruments that help protect the value of property in the event of damaging weather. Government policies that encourage settlement in vulnerable areas such as floodplains (e.g., government-subsidized flood insurance) could be amended to deter such settlement. Government regulations and insurance programs could also be used to deter activities (e.g., by eliminating subsidies) such as farming in vulnerable locales like floodplains, areas prone to freeze damage, or areas where scarce water resources may have a higher value in the future than if they are used for irrigation of low-value crops.

In some circumstances, preventive or protective structures could be built to guard certain areas from flood, storm surge, or hurricane landfall. Coastal barrier walls could be constructed in some valued areas, flood levees may be built to protect low-lying farmland, and other structures such as bridges and causeways reinforced so as to protect them from a wider range of severe weather impacts. Note, however, that there are examples where human tinkering has backfired, such as some levees which,

once broached, exacerbated the Mississippi flooding in the summer of 1993. Municipal and regional water distribution systems could be rebuilt to reduce and/or eliminate leakage, which would conserve scarce water supplies during periods of drought. Changes in water pricing for domestic and industrial users could also encourage water conservation during drought, as demonstrated by pricing conservation measures already adopted in many regions such as the southwestern United States. In areas dependent on groundwater for municipal and/or agricultural uses, regulations for water extraction could promote conservation and even reuse of finite groundwater supplies.

While society may not be able to insulate itself completely from changes in local and regional weather that may accompany a future climate change, human populations do possess intellectual, economic, and physical capacities to manage their vulnerability to severe weather events. Nonetheless, it is important to search for historical lessons as to how well (or how poorly) society may adjust to possible climate-change–induced alterations in regional weather. Society can either continue to repeat mistakes, or learn how and where behavioral and physical adjustments can help manage societal vulnerability to weather phenomena.

VII
The Need for More Research

The complex interactions and feedbacks that occur within Earth's climate system make it difficult to establish just how large the human-induced effects will be or how soon we may be able to detect the climate change unequivocally. It is critical to increase our understanding of the natural variability of the climate system, to build better climate models that more explicitly and more accurately represent weather phenomena, and to reduce uncertainties in predictions of what human activities are contributing to the climate system. In turn there is a great need to be able to better translate what changes in climate might mean in terms of the weather, weather sequences, and extremes that may occur, so that these in turn can be translated into impacts on various sectors of society and human endeavor. In this way, improved strategies for dealing with Earth's ever-changing environment might be effected.

GLOSSARY

Aerosol—Microscopic particles suspended in the atmosphere, originating from either a natural source (e.g., volcanoes) or human activity (e.g., coal burning).

Albedo—The reflectivity of the Earth.

Anaerobic—Occurring in the absence of free oxygen; an example of an anaerobic process is digestion in cattle.

Annual cycle—The sequence of seasons over a full year.

Anthropogenic climate change—Climate change arising from human influences.

Anticyclone—A high-pressure weather system. The wind rotates clockwise around these in the Northern Hemisphere and counterclockwise in the Southern Hemisphere. They usually give rise to fine, settled weather.

Atmospheric chemistry—The science of the chemical composition of the atmosphere.

Atmospheric instability—The growth of small disturbances into large disturbances through internal processes.

Baroclinic instability—An atmospheric instability associated with horizontal temperature gradients such as between the equator and the poles.

Biomass burning—The burning of organic matter from plants, animals, and other organisms.

Carbon dioxide (CO$_2$)—A naturally occurring, colorless atmospheric greenhouse gas. It arises in part from decay of organic matter. Plants take up carbon dioxide during photosynthesis. Animals breathe it out during respiration. Humans contribute to carbon dioxide concentrations in the atmosphere by burning fossil fuels and plants.

Chaos—In a technical sense, a process whose variations look random even though their behavior is governed by precise physical laws.

Chlorofluorocarbon (CFC)—One of a family of greenhouse gas compounds containing chlorine, fluorine, and carbon. CFCs do not occur naturally; all are made by humans. They are generally used as propellants, refrigerants, blowing agents (for producing foam), and solvents.

Climate—The average weather together with the variability of weather conditions for a specified area during a specified time interval (usually decades).

Climate change—Long-term (decadal or longer) changes in climate, whether from natural or human influences.

Climate model—A computer model that uses the physical laws of nature to predict the evolution of the climate system.

Climate system—The interconnected atmosphere-ocean-land-biosphere-ice components of the Earth involved in climate processes.

Climate variation—A fluctuation in climate lasting for a specified time interval, usually many years.

Cold front—A transition zone where a cold air mass advances, pushing warmer air out of the way. Warm air is forced to rise, commonly creating convection and thunderstorms, so that a period of "bad weather" occurs as the temperatures drop.

Composition of the atmosphere—The makeup of the atmosphere, including gases and aerosols.

Convection—In weather, the process of warm air's rising rapidly while cooler air subsides, usually more gradually, over broader regions elsewhere to take its place. This process often produces cumulus clouds and may result in rain.

Cumulus cloud—A puffy, often cauliflower-like, white cloud that forms as a result of convection.

Cyclone— A low-pressure weather system. The wind rotates around cyclones in a counter-clockwise direction in the Northern Hemisphere and clockwise in the Southern Hemisphere. Cyclones are usually associated with rainy, unsettled weather and may include warm and cold fronts.

Dust Bowl era—The period during the 1930s when prolonged drought and dust storms arose in the central Great Plains of the United States.

Dynamics—In climate, the study of the action of forces on the atmospheric and oceanic fluids and their response in terms of winds and currents.

Ecosystem—A system involving a living community and its nonliving environment, considered as a unit.

El Niño—The occasional warming of the tropical Pacific Ocean off South America. Associated warming from the west coast of South America to the central Pacific typically lasts a year or so and alters weather patterns around the world.

Electromagnetic spectrum—The spectrum of radiation at different wavelengths, including ultraviolet, visible, and infrared rays.

Enhanced greenhouse effect—The increase in the greenhouse effect from human activities.

Evapotranspiration—The evaporation of moisture from the surface together with transpiration, the release of moisture from within plants.

Feedback—The transfer of information on a system's behavior across the system that modifies behavior. A positive feedback intensifies the effect; a negative feedback reduces the effect.

Fossil fuel—A fuel derived from living matter of a previous era; fossil fuels include coal, petroleum, and natural gas.

General circulation model—A computer model, usually of the global atmosphere or the oceans; GCMs are often used as part of even more complex climate models.

Glacier—A mass of ice, commonly originating in mountainous snow fields and flowing slowly down-slope.

Global warming—The increasing heating of the atmosphere caused by increases in greenhouse gases from human activities and their "entrapment" of heat. It produces increases in global mean temperatures and an increased hydrological cycle. This phenomenon is also popularly known as the greenhouse effect.

Greenhouse effect—The effect produced as certain atmospheric gases allow incoming solar radiation to pass through to the Earth's surface but reduce the escape of outgoing (infrared) radiation into outer space. The effect is responsible for warming the planet.

Greenhouse gas—Any gas that absorbs infrared radiation in the atmosphere.

Groundwater—Water residing underground in porous rock strata and soils.

Hydrological cycle—The cycle by which water moves and changes state through the atmosphere, oceans, and Earth. Evaporation and transpiration of moisture produce water vapor, which is moved by winds and falls out as precipitation to become groundwater, which in turn may run off in streams or in glaciers into the seas or become stored below ground.

Infrared radiation—The longwave part of the electromagnetic spectrum, corresponding to wavelengths of 0.8 microns to 1,000 microns. For the Earth, it also corresponds to the wavelengths of thermal emitted radiation. Also known as longwave radiation.

Jet stream—The strong core of the midlatitude westerly winds, typically at about 8 to 10 km above the surface of the Earth, in each hemisphere.

Land surface exchange—An exchange of gases from the land surface into the atmosphere or vice versa. The most common is evaporation of water into water vapor.

Little Ice Age—A prolonged cool period, especially in Europe, occurring primarily in the 16th and 17th centuries.

Longwave radiation—See infrared radiation.

Mean—The average of a set of values.

Methane (CH$_4$)—A naturally occurring greenhouse gas in the atmosphere produced from anaerobic decay of organisms. Common sources include marshes (thus the name "marsh gas"), coal deposits, petroleum fields, and natural gas deposits. Human activities contribute to increased amounts of methane, which can come from the digestive system of domestic animals (such as cows), from rice paddies, and from landfills.

Natural greenhouse effect—The part of the greenhouse effect that does not result from human activities.

Negative feedback—See feedback.

Net radiation—The sum of all the shortwave and longwave radiation passing through a level in the atmosphere.

Nitrous oxide (N$_2$O)—A naturally occurring greenhouse gas in the atmosphere produced by microbes in the soil and ocean. Humans contribute to concentrations through burning wood, using fertilizers, and manufacturing nylon.

Nonlinear—Not linear. Linear relationships between two variables can be plotted as a straight line on a graph. Nonlinear relationships involve curved or more complex lines.

Normal distribution—A bell-shaped curve of the distribution of the frequency with which values occur, defined by the mean and the standard deviation.

Ozone (O$_3$)—A molecule consisting of three bound atoms of oxygen. Most oxygen in the atmosphere, consists of only two oxygen atoms (O$_2$). Ozone is a greenhouse gas. It is mostly located in the stratosphere, where it protects the biosphere from harmful ultraviolet radiation. Human activities contribute to near-surface ozone through car exhaust and coal-burning power plants; ozone in the lower atmosphere has adverse affects on trees, crops, and human health.

Phenology—The study of natural phenomena that occur in a cycle, such as growth stages in crops.

Photosynthesis—The process by which green plants make sugar and other carbohydrates from carbon dioxide and water in the presence of light.

Positive feedback—See feedback.

Runoff—Excess rainfall that flows into creeks, rivers, lakes, and the sea.

Scattering radiation—The dispersion of incoming radiation into many different directions by molecules or particles in the atmosphere. Radiation scattered backwards is equivalent to reflected radiation.

Solar radiation—Radiation from the sun, most of which occurs at wavelengths shorter than the infrared.

Southern Oscillation—A global-scale variation in the atmosphere associated with El Niño events.

Stability—In meteorology, a property of the atmosphere, making it resistant to displacements. The atmosphere is stable if a perturbation decays and it returns to its former state. It is unstable if the perturbation grows.

Standard deviation—A measure of the spread of a distribution. For a normal distribution, 68% of the values lie within one standard deviation of the average.

Stratosphere—The zone of the atmosphere between about 10–15 and 50 kilometers above the Earth's surface. Most of the ozone in the atmosphere is in the stratosphere. The stratosphere is separated from the troposphere below by the tropopause.

Temperature gradient—The differences in temperature across a specified region.

Thermal—A rising pocket of warm air.

Thermal radiation—Longwave (infrared) radiation from the Earth.

Transpiration—The giving off of water vapor through the leaves of plants.

Troposphere—The part of the atmosphere in which we live, ascending to about 15 km above the Earth's surface, in which temperatures generally decrease with height. The atmospheric dynamics we know as weather take place within the troposphere.

Urban heat island—The region of warm air over built-up cities associated with the presence of city structures, roads, etc.

Visible radiation—Electromagnetic radiation, lying between wavelengths of 0.4 and 0.7 microns, to which the human eye is sensitive.

Warm front—A transition zone where a warm air mass pushes cooler air out of the way over a broad region. The warm air tends to rise, often creating stratiform clouds and rain as the temperatures rise.

Weather—The condition of the atmosphere at a given time and place, usually expressed in terms of pressure, temperature, humidity, wind, etc. Also, the various phenomena in the atmosphere occurring from minutes to months.

Weather systems—Cyclones and anticyclones and their accompanying warm and cold fronts.

Wind shear—Large differences in wind speed and/or direction over short distances.

SUGGESTED READINGS

Adams, R.M., R.A. Fleming, and C. Rosenzweig, 1995: Reassessment of the economic effects of global climate change on U.S. agriculture. *Climatic Change* 30, 147–168.

Andrews, W.A., 1995: *Understanding Global Warming.* D.C. Heath Canada Ltd., Toronto, Canada.

Dotto, L., 1999: *Storm warning: Gambling with the climate of our planet.* Doubleday, Toronto, Canada.

Gedzelmen, S.D., 1980: *The Science and Wonders of the Atmosphere.* John Wiley and Sons, New York, New York.

Hartmann, D.L., 1994: *Global Physical Climatology.* Academic Press, San Diego, California.

Jones, P.D., M. New, D.E. Parker, S. Martin, and I.G. Rigor, 1999: Surface air temperature and its changes over the past 150 years. *Review of Geophysics* 37, 173–199.

Karl, T.R., R.W. Knight, D.R. Easterling, and R. G. Quayle, 1995: Trends in the U.S. climate during the twentieth century. *Consequences* 1, 2–12.

Karl, T., and K. Trenberth, 1999: The human impact on climate. *Scientific American.* December, 100–105.

IPCC (Intergovernmental Panel on Climate Change), 1996: *Climate Change 1995: The Science of Climate Change.* J.T. Houghton, F.G. Meira Filho, B.A. Callander, N. Harris, A. Kattenberg, and K. Maskell, eds. Cambridge University Press, Cambridge, U.K.

Lamb, H.H., 1982: *Climate, History, and the Modern World.* Cambridge University Press, Cambridge, U.K.

Mearns, L.O., 1993: Implications of global warming on climate variability and the occurrence of extreme climatic events. In *Drought Assessment, Management, and Planning: Theory and Case Studies.* D. A. Wilhite, ed. Kluwer Publishers, Boston, Massachusetts, 109–130.

Mooney, H.A., E.R. Fuentes, and B. I. Kronberg, eds., 1993: *Earth System Responses to Global Change: Contrasts between North and South America.* Academic Press, San Diego, California.

Parry, M.L., 1978: *Climatic Change, Agriculture, and Settlement.* Dawson and Sons, Ltd., Folkestone, U.K.

Raper, C.D., and P.J. Kramer, eds., 1983: *Crop Reactions to Water and Temperature Stresses in Humid, Temperate Climates.* Westview Press, Boulder, Colorado.

Rosenzweig, C., and D. Hillel, 1993: Agriculture in a greenhouse world. *Research and Exploration* 9 (2), 208–221.

Trenberth, K.E., ed., 1992: *Climate System Modeling.* Cambridge University Press, Cambridge, U.K.

Trenberth, K.E., 1996: Coupled climate system modeling. In *Climate Change: Developing Southern Hemisphere Perspectives.* T. Giambelluca and A. Henderson-Sellers, eds. John Wiley & Sons, New York, New York, 63-88.

Trenberth, K.E., 1999: The extreme weather events of 1997 and 1998. *Consequences*, Vol 5, 1, 2–15.

Williams, J., 1992: *The Weather Book.* Vintage Books, New York, New York.

DISCUSSION QUESTIONS

1. Given the large seasonal changes in climate, why are relatively modest changes in climate from one year to the next so disruptive?

2. Weather patterns never repeat exactly, so the variations are apparently endless, yet the range of patterns is limited. Explain this apparent contradiction.

3. People like to blame weather disasters on some cause. Although a disaster can be linked to various weather phenomena, there may not be a single true cause. Explain why.

4. List possible human influences on climate. Note which ones are likely to be global, which are more likely to just be regional, and why.

5. What kinds of weather events most affect human activities? What can be done to prepare for these or ameliorate the effects?

6. What do you think is the value of a climate forecast of the expected weather for the season ahead? How accurate would it have to be to be useful? Given forecasts that might be classed as having (a) some but not much accuracy, (b) reasonable accuracy, or (c) complete accuracy, discuss how this information might be used to economic advantage.

7. Climate changes are under way but are not yet large. People disagree over whether to take action to try to stop or slow human influences on climate. Their views seem to be related to their values and perspectives on what is important now versus how much weight should be given to the future. In a group, put forward your own views on what you think should be done and discuss the range of different views and what factors influence them.

8. People offset risks of natural disasters by taking out insurance. Discuss other ways to mitigate the effects of weather and climate change on various activities.

9. If the climate warms, it is sometimes suggested that crops and plants should just be grown in locations farther north. Explain why this may not be possible (consider especially sunlight, soil conditions, and disease).

10. What does "climate is what we expect, but weather is what we get" mean?

INDEX